BACKYARD WEATHER FORECASTING

How to Be Your Own Weekend Weather Forecaster

Ann M. Beardsley

Dedication

To Wayne Owens—a loving husband to Marian, a handyman extraordinaire to the whole neighborhood, an intrepid pilot, and always handy with a joke. You will be missed.

August 1, 2013

To Harold Robinson——a dedicated teacher, loving father to Amy and David, and with an irrepressible sense of humor. You too will be missed.

August 17, 2013

Acknowledgments

Thank you to my beta readers, Jackie Ogburn, Becky Lallier, and Ellen and Dooley Banks. Thanks to NOAA for many of the weather graphics. And, as always, thanks to Elliott Walsh for support, advice, food, time, and, well, just about everything.

Thanks also to my readers, especially those of you who take time to send me emails, reviews, and a general acknowledgment of my existence. Without you, this would simply be pixels and bytes of unreadable, unrecognizeable data in a flood of the same.

Contents

INTRODUCTION

Weather is a fascinating subject that can scarcely be covered in one short book. My goal is to give you enough of an overview—without getting too technical—to enable you to forecast with reasonable accuracy the weather you'll be getting on a short-term basis—or, roughly, within three to five days.

Your TV, Internet, newspaper, and other weather forecasters do a tremendous job for the areas they cover. One problem, though, is that weather is location dependent—just like that old real estate joke about location, location, location. Where I live, we get much less rain than the folks just six miles away on the other side of the highway. Ninety miles north of us has seen far more severe weather than we have. Yet the same weather forecasters cover both areas—and chances are, their forecasts were spot-on for at least one small part of the geographic area they cover. But I don't really care about the weather ninety miles away—I want to know what it's going to do here, in my own piece of the world. Do I need to water my garden this morning, or will it rain here this afternoon? Will the snow hold off for my daughter's wedding this week or should I put my backup plan of renting the Kiwanis Hall into effect? Can I go boating tomorrow or will I get caught in a thunderstorm? (We in the United States tend to be a self-absorbed society, but in this one respect I think we're fully justified...)

Naturally, nothing is guaranteed. No one is 100 percent accurate 100 percent of the time. There are far too many variables to be able to know precisely that the drizzle will begin at 10:08 a.m. and end at 1:30 p.m., or that the wind will be exactly 27 miles per hour from the southwest at 190 degrees. Still, you can predict trends—and that's what we're looking for here. You still won't be right every time. You might think it's going to rain, and it does—right next door, missing you by scant inches. It happens—after all, there's got to be an edge to the rainfall somewhere, and why not there?

We'll also go over how to interpret the weather charts you can find on the Internet on some of the publically available free sites that can be specified to your location.

You'll be able to make plans for a picnic this Saturday with confidence that you won't need a tent—or conversely, you'll have a tent in the back of your truck and be the Scout (Girl or Boy) who's always prepared.

You'll notice throughout the book that I use "weasel" words: tend to, generally, usually, probably. This is not because I don't want you showing up at my house to complain because your daughter's graduation from medical school ceremony was rained out (although that's certainly true), but because there are very few absolutes in weather forecasting. And those absolutes are rather useless:

- Eventually it will stop raining.
- Eventually it will rain again.
- And so on.

Every so often, you'll see text in a box, like this. The information in the box is not critical to forecasting the weather but includes interesting scientific or technical information on other aspects of weather. Most of the time, it will include additional resources or websites where you can explore in depth many of the fascinating topics that make up weather forecasting. Sometimes it will include a joke.

Either way, If you just want to get on with forecasting, feel free to skip over the boxes.

I can think of only one that's even middling useful:

- A change in weather is always accompanied by a change in wind direction before the weather arrives.

Helpful, huh? Weather forecasting is a science of trends, and trends are weaselly by definition.

A pet peeve of mine is when forecasters refer to "good" or "bad" weather. I think "fair" and "wet" weather are more accurate, as there are many times (mostly when I don't want to go outside and water my garden) that a little "bad" weather would be a very good thing indeed. Without the "bad" weather, we'd live on a very dry planet—kind of like the moon, and we all know how much intelligent life is found on the moon.

1 THE SCIENCE OF WEATHER: BIG PICTURE BASICS

With weather, everything works together. The wind, the temperature, the precipitation, the atmospheric pressure, and the clouds all move together in an intricate dance of ever-changing positions. Change one small item, and the rest must dance to its tune. In chaos theory, this is what's known as the butterfly effect. The name was coined by Edward Lorenz in 1961 but the idea had been around for a long time. The theory is that a butterfly wing's tiny change of wind in, say, the Philippines will escalate, morph, and modify the weather such that a tornado forms (or doesn't form, or isn't as strong, etc.) sometime later in, perhaps, Oklahoma.

Weather occurs on a large scale—for example, a large cold air mass (whose leading edge is called a cold front) might cover most of Canada before it heads southeast. It pushes out ahead of it a large warmer air mass, whose trailing edge (in this case) is usually called a warm front. Fronts can be as long as a couple thousand miles, or as short as a few hundred miles. In fact, some scientists believe there is a single continuous front going all around the globe, twisting and turning in all directions. When two air masses collide, wet weather often results. The sinking air of a cold (and often drier) front will sneak in under the warmer (and often more moist) air of a warm air mass, causing the air to rise and fall, twirl and spin, not unlike milk in your coffee.

Weather tends to move from west to east in the United States—but this is by no means a sure thing. Picture the big storms coming from the Gulf Coast and heading northeast, or the snowstorms coming from Canada and heading southeast. Generally, the weather has a westward component, even if smaller sections of it—such as the northern section of a hurricane in the northern hemisphere—might have winds blowing from the east. This tendency to move from west to east is called zonal flow, and is related to the Earth's rotation around the sun.

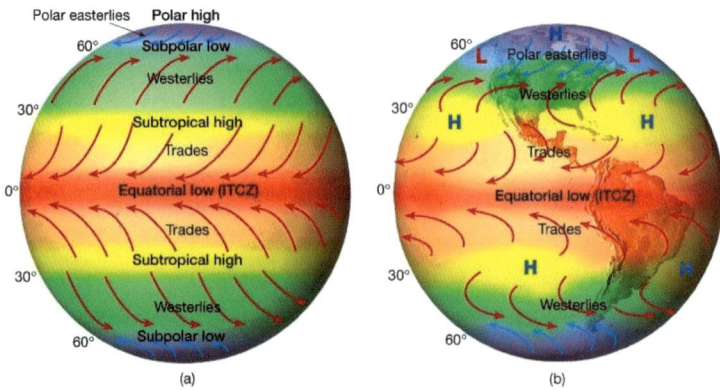

(a)   (b)

Figure 2.1. (a) Idealized winds generated by pressure gradient and Coriolis Force. (b) Actual wind patterns owing to land mass distribution. Notice the westerly flow across the United States. (Photo courtesy of *The Atmosphere*, 8th ed., Lutgens and Tarbuck, 2001.)

Weather satellites help us see these larger systems at work, as does the weather reporting equipment that monitors the pressure, wind speed, wind direction, and temperature at various levels throughout the atmosphere. Satellites contributed greatly to our understanding of the weather patterns associated with the *warm* El Niño (named "The Child" because they arrive around Christmas time) and corresponding *cooler* Las Niñas. These systems are part of oscillations that move up, down, and across the oceans in patterns that we are only beginning to understand.

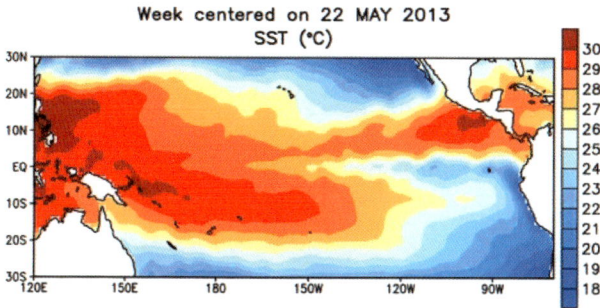

Figure 2.2. Sea surface temperatures showing location of El Niño. La Niña would be a similar representation but of cooler waters. (Photo courtesy of NOAA.)

Better understood is the greenhouse effect, which is the radiation from the sun that hits our planet, is absorbed by the Earth's surface, and re-radiated back to space. Part of that thermal radiation hits our atmosphere on its way back to space but gets sent back to the surface, bouncing back and forth between Earth and atmosphere. As the air bounces back and forth, greater and greater amounts of greenhouse gases absorb and re-radiate energy back toward Earth, making us warmer. By burning fossil fuels and wood, we increase the amount of gases (mainly carbon dioxide, CO_2), which causes us to be even warmer. Most of the fossil fuels (coal and oil) are used in electricity production.

This short digression helps explain why we cannot necessarily rely on climatological data from generations past. We appear to be increasing our temperature at an increasingly higher rate, which will at some point in the future have dire consequences for the planet (decreased land masses, more desert areas, and an increase in severe storms because of temperature and precipitation extremes). It may not affect your weather this weekend, but it may help explain some of the disparities between your grandparents' weather and yours.

The weather in your neck of the woods is related to weather all across the planet. In that sense, we are a truly global people. Your weather here may have been precipitated (pun intended) by weather on the far side of the planet. For example, the Sahara Desert is far, far away. But storms form near the intercontinental tropical zone (ITCZ) and minute sand particles become the condensation nuclei around which raindrops form. The more sand caught up in the storm, the greater the potential for rain in the presence of water vapor. When all those sand particles are caught up in a storm that moves northward into the ITCZ, tropical depressions form, strengthen, and some become hurricanes—which then disperse all those sand particles (in the form of raindrops) all over the eastern Atlantic. The next time it rains on you, think desert sand.

2 WIND

The short and sweet definition of wind is "air that is moving." Air moves due to the heat of the sun. Remember in fourth-grade science you probably learned that "warm air rises, cold air sinks"? The same holds true everywhere on Earth. At the equator, which is closest to the sun and therefore warmest, air rises. The Earth radiates heat too, and as all that air rises away from warmth of the Earth, it cools. When it reaches a point of (more or less) equilibrium—when it has the same temperature as the surrounding air—it spreads out and, when it cools, it sinks.

Air also moves from areas of high pressure to areas of low pressure, filling in the low places just like when tide water rushes in to fill the holes made by your feet. And at the same time the air is rising and falling, seeking out those areas of low pressure, the Earth is rotating, causing the air to spin in circular patterns.

But the surface of the planet isn't flat—we have mountains, oceans, deep valleys, deserts, trees, and tall buildings. Those areas get in the way of the wind, causing friction (not to mention cooling as air moves up a mountain, warming as it sinks). Every place on Earth has its own individual peculiarities that make it unique in terms of weather. Picture the wind charging down the streets of Chicago, channeled between the buildings (hence the name Windy City), or the soft gentle land breezes that occur in the evenings along the coast.

There are various forces that act on air, causing some degree of wind. Remember that air is simply a set of gases that tends to move in a crowd, somewhat like teenagers. Each gas thinks it is unique (well, if gas could think), yet it moves in a fairly predictable pattern (just like cheerleaders, jocks, geeks, and social misfits):

- *Pressure gradient force:* Air tends to move from high-pressure regions to low-pressure regions.
- *Coriolis force:* Air tends to move in a clockwise pattern in the Northern Hemisphere (counterclockwise in the Southern Hemisphere), and is negligible at the equator and the poles.
- *Frictional force:* Air tends to be slowed down or speeded up depending on what it encounters.
- *Centrifugal force:* Air tends to be forced outward in a rotation. (Imagine a bucket of water on a string, where the water stays in the bucket as you twirl around.)
- *Gravitational force:* Air tends to move downward and perpendicular to the Earth's surface.

Because wind is such an important component of weather—and because it can do so much damage—we have assorted names for recurring types of wind (hurricanes and tornadoes are covered in chapter 9), ranging from Abroholos to Zondo. These are all almost local winds, and if you've heard their names floated about, it might be relevant to you. See http://library.thinkquest.org/C001472/en/development/types.content.html for a listing of many of these local winds, with corresponding maps of the locations they affect. If these do affect you, make note of it on your worksheet in chapter 14.

Wind is measured by an anemometer. In its absence, a land Beaufort wind scale chart can give you a rough guess of wind speed (an equivalent marine Beaufort wind scale chart is used over water). A copy of the chart is available at the end of this chapter.

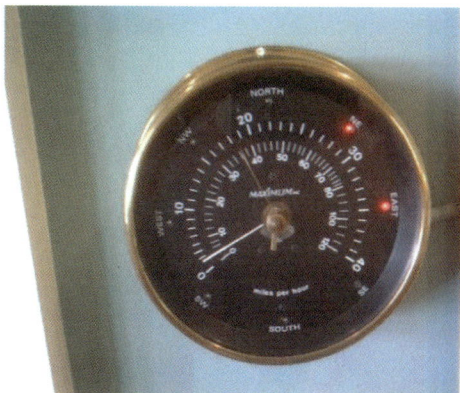

Figure 3.1. Analog anemometer.

Gusts are short bursts of stronger wind. These often precede frontal passage and can cause significant damage. Unless you're lucky enough to be watching for wind speed just when one happens, you might miss it if you're relying on visual clues. Many anemometers have a wind gust detector that you can set to zero after a storm. During a storm, each stronger gust pushes the needle a little further. This one here shows a peak wind gust at about 17 mph, but a current wind from the east-northeast of less than 2 mph.

Clouds can also give you an indication of wind speed. Sharp edges on a cloud tend to indicate strong wind, particularly rotor clouds (which are usually formed on the eastward side of mountains and indicate turbulence; small-plane pilots are warned to steer clear of these). Puffy, cotton-ball edges on clouds mean that no strong winds surround the cloud. On thunderstorms, you'll often see an anvil shape forming at the top of the cumulonimbus cloud, which means that the cloud tops have reached the tropopause, the boundary of the Earth's atmosphere (generally around 30,000 feet at the poles and 56,000 feet at the equator). As a general rule, however, the sharper the cloud's edges, the stronger the wind.

But don't count on clouds to tell you which way the wind is blowing every time. Significant sheer in the atmosphere can cause one layer of clouds to move one direction, and a layer of clouds just below to move in the opposite or a different direction.

In the absence of other prepositions, wind direction is always telling you which direction the wind is from—in other words, a northerly wind is flowing from the north. It's important to keep that in mind when forecasting wind direction. One rule of thumb is to take the weather from 200 miles upstream: what they have today, you'll have tomorrow (not allowing for any local geographic variations).

When forecasting winds, your upstream component is critical. Wind doesn't simply appear with no warning, as it always comes *from* somewhere, though its speed might catch you by surprise. Typical wind direction varies by season, by location, and even by time of day. This will take a few days of observing, days on which no serious weather occurs. Where I am, the prevailing wind is northwesterly at about 1–2 mph. It tends to pick up near sunset, perhaps 4–5 mph. If it changes, then I can generally assume a change in the weather will happen within the next couple of days. I also know that I am on the East Coast, and that weather generally arrives from the northwest, which means that I don't usually expect a lot of moisture in the air. However, in the warmer months, we get a lot more weather from the southwest—from the Gulf Coast. That air is more humid and warmer, and generally more unstable—so I have a better chance of rain and thunderstorms.

A shift in wind will mean a shift in weather. For me, a shift from northwesterly dry air to southwesterly moist air means I'll likely have higher humidity, a greater chance of rain, and warmer temperatures.

See the Wind/Barometer table below for a general weather pattern in the United States. Note that your location may be different. If, for example, you live in Buffalo, where 200 miles upstream is Lake Ontario and its vast supply of moisture, you're going to get a lot more wind and rain/snow than someone who lives 200 miles downstream from you. If you live in a green valley on the eastern edge of a desert, the same thing applies: you'll get hotter, drier air than the people who live 200 miles downstream.

Why are there stock market analysts?

To keep weather forecasters from feeling badly about themselves.

Keep in mind also that diurnal variations (that is, ones that occur on a daily basis) will have an effect. In the South, many afternoons at 4:00 there'll be a brief shower—even though the wind is calm, the pressure is steady, and there's no front pushing through. Why? We'll get to that later, but in general, thunderstorms need two things: moisture in the air and something to trigger it. The high afternoon temperatures do both, as warmer air holds more moisture and sets off enough instability with the rising and cooling air to produce a thunderstorm.

If it were just wind you needed to watch, I'd leave you with this handy chart and we'd be done. But wind is just one component of your weather forecast; let's move on to pressure.

Forecast tips:

On calm, clear nights, the warmer air nearest the surface of the Earth rises and the Earth cools off. But on a windy clear night, as the air is churned by the wind, the wind allows the warmer air to be pushed back to the surface, and temperatures don't always drop as quickly. When forecasting temperatures, then, you might predict a slightly warmer night if you expect winds.

Light and variable winds—those less than about 7 mph that vary in direction—have no importance to weather forecasting. They are simply the result of minor wind variation. These winds will circulate through all directions of the compass without regard to any weather patterns (though they are still subject to planetary forces such as rotational, gravitational, and centrifugal forces).

Land Beaufort Wind Scale

Beaufort #	Description	Knots	Km/Hr	MpH	Visible Signs
0	calm	0–1	0–2	1	smoke rises straight up
1	light air	1–3	3–6	1–3	smoke drifts
2	light breeze	4–6	7–11	4–7	wind felt on face, leaves rustle
3	gentle breeze	7–10	12–19	8–12	flags flap, twigs move all the time
4	moderate breeze	11–16	20–28	13–18	papers blow, small branches move
5	fresh breeze	17–21	29–38	19–24	small trees sway
6	strong breeze	22–27	39–49	15–31	large branches move, wind whistles
7	near gale	28–33	50–61	32–38	whole trees sway
8	gale	34–40	62–74	39–46	twigs break off, gale warnings on radio
9	strong gale	41–47	75–88	47–54	large branches break, some damage, roof tile damage
10	storm	48–55	89–102	55–63	trees uprooted; major damage
11	violent storm	56–63	103–117	64–75	DANGER— TAKE SHELTER
12	hurricane force	64+	118+	75+	DISASTROUS

Note: Take local conditions into account when judging by trees. Where I live, pine trees planted some forty to sixty years ago abound (and also make it hard to see more than bits and pieces of the sky). The trees are older, more susceptible to damage, and sway in fresh breezes. Limbs can fall off during even strong breezes. Be alert to conditions around you. If you live in the desert where trees are few and far between, perhaps you can judge by movement of sand, a flag, or even a string on a pole. If you live near the ocean, maybe the fronds of the palm trees can tell you or the size of the waves.

3 PRESSURE

Weather is caused by moving air, and air moves to areas of lowest pressure—just as water sinks to the lowest spot (the motion is the same, even though the dynamics may be different). So if you have an area of low pressure downstream, the upstream air—which is an area of higher pressure—is going to try to flow right over you as it rushes toward the low-pressure area.

Combine this tendency with the fact that warm and humid air masses have a lower surface pressure than cold, dry air masses and you have a lot of cooler air moving toward an area of warmer air. One component of the movement of air concerns the difference in their temperature—a cold body (of higher pressure) is going to rush toward a warm body (and its corresponding low pressure). The greater the difference between the two, the faster the air moves. (I think I'd rush toward a warm location on a cold wintery day too, but that's not to say that air masses actually think.)

There are exceptions, but in general, that's the way pressure works. (Maybe this is why one image that comes to mind for a vacation is to be sunbathing on a beach—an area of warm, low pressure to which all the cold, high-pressure folks are flocking.)

Pressure is measured in inches of mercury (now mostly digital measurement) and is always converted to sea level pressure (29.92") when comparing pressures at two different locations. Think of this as a measure of a 1" column of air directly on top of your head, going all the way up about 20 miles. If you are in Denver, you're going to have a lot less air on top of your head than if you are in Death Valley, so it's going to "weigh" less in Denver. Atmospheric pressure is also measured in millibars, and 1013.25 mb = 29.92 inches of mercury (and for the technically minded among you, = 14.7 psi).

This is a transition time for barometers. Mercurial barometers—the ones that measure in inches of mercury—are more accurate than those that measure in millibars and are usually the standard on which the others are based. There are fewer and fewer mercurial barometers, however, and it is difficult (if not impossible) to send mercurial barometers up in a weather balloon, so the tendency these days is to use digital equipment and report in millibars that are converted to sea level pressure (otherwise the highs and lows might simply indicate geographic valleys and mountains and not their air equivalents).

To make matters even more confusing, the decimal points in millibar readings are eliminated. A pressure of 1013.25 is reported as 1325; one of 992.6 is reported as 926. An isobaric (iso=same, so lines of equal bars, or pressure) chart is an easy way to pinpoint areas of high and low pressure, but requires atmospheric data to show. An example of an isobaric line chart is shown below.

As that relatively cold mass of air moves toward the low-pressure area, it tends to rotate (remember the Coriolis force?) in an inward, counterclockwise manner (in the Northern Hemisphere). As the cold air fills in the "sinkhole" (which is anywhere from 500 to 1200 or so miles across), it pushes the warm air up. The warm air, obligingly, wants to rise. In the summertime, this brings in the cooler air on the surface—unless it started in the tropics, in which case it brings, alas, even warmer air—but that's one of the exceptions. In the winter, it brings higher winds and colder temperatures. If the low-pressure system moves directly over you, you can have strong winds and considerably cooler weather.

At the same time, where's the warm air to go? Some of it mixes in with the cold air, but some forms its own high-pressure system, this time ranging from 2000 to 3000 miles across. When high-pressure areas are overhead, you've generally got fair weather; winds are calmer in the center, spiraling outward in a clockwise direction. High-pressure systems bring little temperature change and are windy only towards the outer edges (just like that bucket on a string, centrifugal force).

You can track a pressure system by watching which way your wind blows. A high-pressure system, which is going clockwise around you, will begin coming from a southerly direction as it begins to leave your area. As the low-pressure system—and its attendant cooler temperatures and stronger winds—moves into your area, the winds will become more easterly (that is, from the east)—even though the entire low-pressure system is moving slowly in an eastward direction.

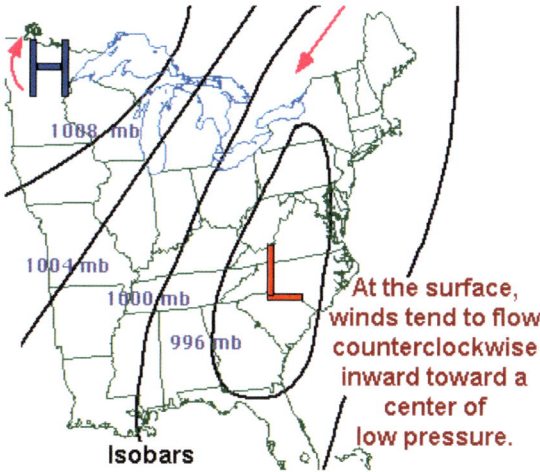

Figure 4.1. Isobaric chart showing wind direction. (Drawing courtesy of NOAA.)

It would be rare, however, for the center of the high-pressure system to move directly over you, followed by the center of the low-pressure system, so you need to be alert to the change in wind direction based on where the pressure systems are. Keep this basic truth in mind: A change in wind direction leads to a change in weather, and probably sooner rather than later.

Pressure is all relative. An area of low pressure is only low relative to the air around it; likewise, an area of high pressure is high only relative to the air around it. Some of these areas of low or high pressure are found high in the atmosphere, and only barely drop to the surface.

Forecasting tips:

Rather than concentrate on numbers, look for trends in pressure. A rising pressure will generally mean better weather, and a falling pressure will generally mean wind with rain or snow. (And yes, this means keeping track of pressure and recording it, unless you have a really good memory.)

Next we'll talk about the fronts and air masses that help define the "edges" of the pressure systems.

Wind/Barometer Table

Wind direction	Barometer reading (inches of mercury)	Type of weather indicated
Southwest to northwest	30.00–30.20, steady	fair weather, not much change for 1–2 days
Southwest to northwest	30.10–30.20, rising rapidly	fair weather, rain within 1–2 days and warmer
Southwest to northwest	30.20, falling rapidly	cold, clear, followed by rain and warmer weather
Southwest to northwest	30.20+, steady	continued fair weather with not much change
Southwest to northwest	30.20+, falling slowly	fair weather for 2 days, slowly rising temperatures
South to southeast	30.00–30.20, falling rapidly	rain within 24 hours, warmer temperatures
South to east	30.20+, falling rapidly	rain within 36 hours, warmer, higher winds
South to east	30.00–, falling slowly	rain with 18 hours, continuing 1–2 days
South to west	29.80–, falling rapidly	severe storm, wind/rain (snow) imminent, colder
South to northeast	30.00–, falling rapidly	high winds, rain, but within 1–2 days, clear and cold
Southeast to northeast	30.00–, falling slowly	rain will continue for 1–2 days
South to west	30.00–, rising	clear and cold within 12 hours
East to north	29.80–, falling rapidly	severe gales, heavy rain (snow); cold wave in winter
East to northeast	30.10, falling rapidly	summer: rain within 12–24 hours; winter: rain/snow, windy
To the west	29.80–, rising rapidly	colder, clearer

Note: These atmospheric pressure indicators are in inches of mercury. A conversion table can be found at http://www.srh.noaa.gov/ffc/html/prescalc.php. Your pressure must be converted to sea level pressure, which requires an additional calculation found at the same link. A shortcut: If you live at elevation, decrease pressure by 1 inch for every 1000 feet of elevation. Also note that you should regularly compare your instrument with the National Weather Service's reported pressure to help determine any local variation.

4   FRONTS

Air masses are particles of air that travel in a group, just like bus passengers on their way to Orlando. These air masses tend to be made up of similar particles (humans) heading slowly toward an area of lower pressure (Disneyworld). Air masses are huge—perhaps 1000 miles across and a couple miles high—and tend to be high-pressure systems (relative to the lower pressure of Disneyworld). These groups of particles can be from the polar regions (i.e., chilly Northerners who long for warmth) or the tropics (Southerners who are searching for a bit of air conditioning), and they travel with particles from their regions (their extended families). They can be stable (mom) or unstable (Aunt Betty) in the upper atmosphere and can further be subdivided between warm core (cuddly kids) and cold core (distant relatives).

The transition zone between two distinctive air masses is a front, and there are four types of fronts:

- Stationary fronts tend to hover in place, and whatever air they bring with them tends to stay for a while.
- Warm fronts are where warm air replaces cooler air. They are slow moving; if there's weather associated with it, it tends to be more widespread, and the warm air simply remains on top of the colder air (leading to more cloud cover).

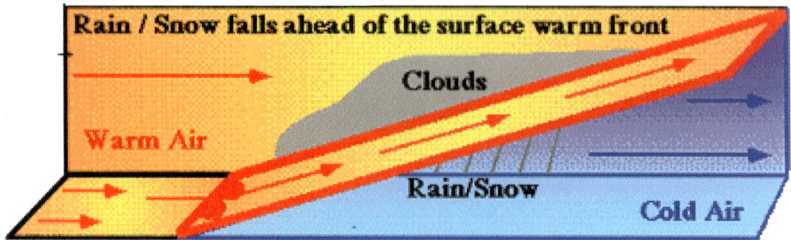

Figure 5.1. Warm front. (Drawing courtesy of NOAA.)

- Cold fronts push a warm air mass out of the way, and the warmer air mixes energetically with the cold front, causing a quick rise in air (leading to towering clouds)

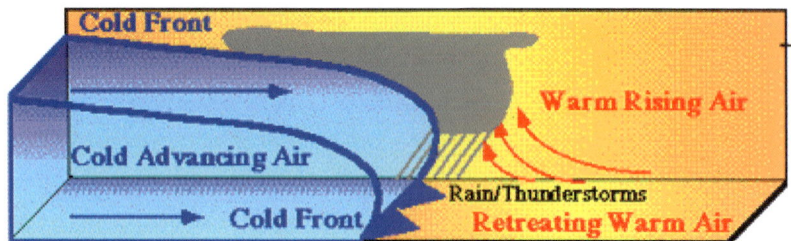

Figure 5.2. Cold front. (Drawing courtesy of NOAA.)

- Stationary fronts are ones that, as the name implies, don't move (although the air masses within them can move). These can linger for days, and a stationary front in winter can precipitate significant blizzards. Mountain ranges and geographical boundaries can sometimes act as a barrier, causing fronts to stall and act like stationary fronts.

- Occluded ("hidden") fronts are usually generated by a low-pressure system where the cold front has overtaken and merged with the warm front. You can get all sorts of weather from one of these, generally back and forth between weather typical of warm and cold fronts, and they can be very slow moving.

There are also lines of discontinuity, like drylines, squall lines, and so forth, but we won't go into those here.

Fronts (or boundary zones) don't come with a handy red or blue line moving across the land like we see on the weather channel. We have to find them, and they're invisible. They do, however, leave clues:

- Wind: a change in direction of 30 degrees or more in a counterclockwise direction.
- Temperature: a large temperature difference between two points (usually greater than 10–15 degrees, but not always).
- Weather: there's generally wet weather near a front; thunderstorms or showers represent a cold front, and steady rain or drizzle a warm front. Fog sometimes occurs ahead of a front, and anything goes on a stationary front.
- Temperature and dewpoint differences: Drier air is behind a cold front, but at the front the temperature and the dewpoint temperature are nearly the same. The dewpoint temperature is the temperature at which the air can no longer "hold" all of the water vapor that is mixed with it, so the water vapor "overflows" the ability of the air to hold onto it, and it condenses into water or other precipitation.
- Pressure: a sinking pressure changes to rising with the crossing of a cold front. Warm fronts are harder to discern via atmospheric pressure, though they are always located in a trough (upper air and low-pressure line of decreased pressure).

Putting this all together leads to another handy chart.

Weather Changes Due to Frontal Passage

	Warm Front		Warm Sector	Cold Front	
	Approaching	Passing		Passing	At the Rear
PRESSURE	Steadily falling	Stops falling	Little change or falling	Sudden rise	Rising steadily
WIND	Veers from S–SW and rises	Veers with possible increase in speed	Steady, possibly backing as cold front approaches	Sudden veer from S–SW or NW with squalls	Velocity tends to decrease, but steady in direction
TEMPERATURE	Rising slowly	Slight rise	Little change with relatively high temperatures	Sudden drop	Little change but tendency to fall
SKY	Becoming overcast Ci, Cs, As, Ns[‡]	Ns and fractal St[‡]	Overcast with St changing to Sc[‡]	Cb[‡]	Cb, As, and Cu[‡], with blue sky
PRECIPITATION	Continuous from drizzly to heavy rain or snow	Rain stops but may be slight drizzle	Intermittent slight rain or drizzle, possibly fog	Heavy rain, thunder, possibly hail	Possibly continuous heavy rain changing to heavy showers
VISIBILITY	Deteriorating	Poor	Poor near cold front	Great improvement	Excellent except in showers
HUMIDITY	Increasing	Rapid rise	Very high	Rapid fall	Fairly low, but variable in rain

Source: Chart modified from Huo-Jin Huang, *ATMS 103 Introduction to Meteorology Supplementary Material*, UNC Asheville (January 2007).

[‡]See chapter 6 for explanation of cloud abbreviations.

5 CLOUDS

There are at least twenty-seven distinct types of clouds, which are divided into low, middle, and high clouds. Each cloud can tell you something about the weather that surrounds it. Fog is simply a cloud that is found at the surface of the Earth.

In addition to the altitude indicators—low (cloud base up to 6500'), middle (cloud base up to about 14,000'–20,000), and high (cloud base from middle on up to around 35,000'), we also have cloud types: cumulus (puffy or heaps), stratus (flat, soupy layers), and cirrus (wispy, curly).

Clouds need water vapor to form as well as condensation nuclei—a tiny particle around which the water can aggregate. Ice-forming nuclei help water vapor form ice crystals. Clouds also need one more thing: rising air. In southern and southwestern states, the hot afternoon air rises, causing puffy cumulus clouds to grow to thunderstorm size when there is enough water vapor in the air.

Pictures say a thousand words, and I recommend looking at various websites, particularly that of the National Oceanic and Atmospheric

Administration, http://www.crh.noaa.gov/lmk/?n=cloud_classification. On paper, clouds can be identified by their abbreviations, and the chart in chapter 5 listed some of the following:

- o Cumulus clouds, which can be found at all three levels.
    - o Cumulus (Cu): low, puffy clouds seen typically on a summer day
    - o Stratocumulus (Sc): rows of low, puffy clouds, with dark bottoms
    - o Altocumulus (Ac): mid-level cumulus clouds, darker in color
    - o Towering Cumulus (Tc): tall cumulus that can grow into cumulonimbus
- o Cirrus clouds, those formed high in the atmosphere
    - o Cirrus (Ci): wispy, also called mare's tails.
    - o Cirrocumulus (Cc): tiny puffy clouds, "mackerel sky"
    - o Cirrostratus (Cs): layer of flat white clouds
- o Stratus clouds, those found in layers
    - o Stratus (St): low-level layers of darker clouds
    - o Altostratus (As): mid-level stratus clouds
    - o Nimbostratus (Ns): low-level rain clouds; these form in the mid-level and "grow" downward
- o Cb: Cumulonimbus: Thunderclouds, which have low bases and can reach all the way through our atmosphere

The types of clouds and how quickly they move can give you a fair idea of what weather is coming.

Forecasting by Clouds

Cirrus	A change in the weather within 24 hours
Cirrus - Few	Fair weather is coming
Cirrus - Increasing	Warm front approaching
Cirrus changing to cirrocumulus	A cold front and associated storm is coming
Cirrocumulus alone	Fair but cooler weather
Cirrostratus	Rain (or snow) is likely within 12–24 hours; the thicker the clouds, the more likely the precipitation.
Altostratus - Increasing	Warm front approaching, rain likely
Altostratus - Decreasing	Front has passed, skies clearing
Altocumulus	Generally fair weather clouds, but thunderstorms in the afternoon if seen on a humid morning
Altocumulus and cirrocumulus	Precipitation within 12–24 hours when also accompanied by cirrostratus
Cumulus	Fair, dry weather, but depending on wind, they can move out and make way for greater cloud development.
Cumulonimbus	Thunderclouds, extreme weather possible. The anvil points in the direction the storm is moving.
Stratus	Drizzle, if there is enough water vapor in the air; clearing likely if fog develops overnight
Stratocumulus	Fair weather by themselves, though they can turn into

	nimbostratus. By themselves, however, dry weather and only light precipitation, if any. A sky full of stratocumulus means little difference in temperature between night and day.
Nimbostratus	When thick enough to block out the sun, will often cause rain or snow that can last for a many hours, often leading to flooding.

Clouds can also be described by how much of the sky the individual types of clouds cover. A cloud (or groups of clouds) of the same type is described as "scattered" when it covers less than half of the sky; it is called "broken" when it covers up to 9/10 of the sky; and a sky is "overcast" when it covers at least 9/10 of the sky. (Overcast skies tend to be stratus layers; stratus layers tend to come with warm fronts; warm fronts tend to longer-lasting precipitation. However, a cumulonimbus cloud directly overhead can block most of the sky and therefore be overcast, but it tends to move quickly, leading to a broken or scattered sky.) It is possible to have many types of clouds in the sky at once, though there generally tend to be four or fewer layers.

During the day, clouds reflect about the sun's heat back into space— maybe 20 percent of it. Another 20 percent or so of the sun's heat is absorbed by clouds, water vapor, and atmospheric gases. Stratus clouds are especially good at reflecting heat back to the sun. The clouds keep us from getting the full heat of the sun, and during the summer that's a particularly good thing. At night, however, a cloudy sky usually will bring slightly warmer temperatures because the heat from the Earth's surface is kept close to the planet by the "blanketing" effect of the clouds.

6 PRECIPITATION

Precipitation—whether rain, snow, sleet, hail, freezing rain, drizzle, fog, or anything else—begins in a cloud, but not just any cloud. The high-level clouds (Ci, Cs, Cc) and altocumulus do not produce precipitation.

Like clouds, precipitation is formed by rising air and water vapor. Late afternoon thunderstorms can be caused simply by higher air temperatures, as long as there is enough moisture in the air. As the air heats up, it rises. When the cloud droplets become large enough, they fall back to the Earth as precipitation. Other forms of precipitation—and we'll call it rain here, although it applies to most types—need a stronger trigger. Often, a front—whether warm or cold—can provide that forcing mechanism. Air on one side of the front is moving in one direction, while air on the other side is going in a different direction. Where they converge, the air must go somewhere—so it goes upward. As the air moves upward—again, if there's enough water vapor—the condensation nuclei (microscopic bits of dust, pollution, smoke, etc.) attract cloud droplets, which combine, coalesce, and then fall.

Rain often forms but then dries out before it reaches the ground. These "rain" showers are called virga, and you'll often see them in or near thunderstorms as dark shafts of raindrops that don't make it all the way to Earth. They've been implicated in the formation of microbursts, as their evaporation from liquid to vapor removes heat from the air beneath the cloud. As that cold air sinks—suddenly and explosively—it causes a dry microburst on the ground below. Virga can likewise be a source of condensation nuclei for nearby thunderstorms, magnifying the amount of rain. Virga is almost impossible to forecast.

There are numerous factors to keep in mind when forecasting rain, but most are a trend of what's happening upstream. A lot of other information is available online, based on Skew-T charts and speed/direction of winds aloft. Knowing that the winds high in the atmosphere are light means that your rain will be more slow moving and take longer to clear your area. A Skew-T chart can give you upper-atmosphere indications of rainfall. But even without these more advanced indicators, you can still have a pretty good idea of whether or not it will rain by watching the other markers:

- What clouds are moving in?
- Is it raining upstream?
- Is the pressure rising or falling?
- Is there a front approaching?
- Is the weather heading your way warmer or colder?

Rain that lasts for a while is generally initiated in altostratus and nimbostratus clouds. Showers (which are by nature intermittent, meaning they stop and start) are more likely from stratocumulus and cumulus, while heavy showers tend to come from cumulonimbus.

Stratus clouds generally produce drizzle and snow (depending on temperature and the freezing line), and snow is possible from all types, though rarely from cumulus and cumulonimbus. Freezing rain (or freezing drizzle) is most often found in a narrow area on the cold side of a warm front when surface temperatures are at or just below freezing, though it can be found with cold fronts and with high-pressure systems when they encounter polar air masses. Ice on the ground is caused when the supercooled drops hit the surface (or the power lines, tree branches, homes, etc.) and freeze on impact, forming a thin layer of ice. A prolonged rain passage will create thicker ice as the droplets build up.

Hail is produced by cumulonimbus and is caused by the up and down drafts within the cloud, and the subsequent freezing, thawing, refreezing, rethawing, etc. The longer the hail is aloft in the storm, the larger it gets. Sleet (or ice pellets) are like frozen raindrops—they bounce when they hit the ground.

When the outside air temperature is near freezing, small temperature changes can make the difference between rain, sleet, freezing rain, or snow, and so it is hard to forecast which type of precipitation will arrive. Generally, however, ahead of a warm front, your first indication is snow, then sleet, then freezing rain, followed shortly by the warm front itself.

Forecasting the amount of precipitation is difficult, but it boils down to: the larger and slower moving the clouds, the greater the precipitation. Blizzards, for example, have winds of at least 35 mph (even though the storm itself isn't necessarily moving at the same speed), with air temperature around 20°F, but they can happen with colder temperatures. Cumulonimbus clouds produce intermittent heavy rainfall, while alto/nimbostratus produce steady rainfall—and either can produce flash flooding.

Because predicting the amount of rainfall is so very difficult for an on-the-ground amateur forecaster, I recommend you head to http://radar.weather.gov/ for rainfall estimates. The radar images provide a moving image of the storm as it heads to you if you choose the "Loop" option. The darker the red, the heavier the rain predicted (just click on the dot closest to you; it does require Java/Javascript).

A word for pilots and sailors: Aviation weather and marine weather have their own particular idiosyncrasies. A lot of this may apply, but use common sense and rely on the experts. Please, do not put your life in danger.

7   TEMPERATURE

Because temperature is based on so many factors—wind (thermal advection), clouds, pressure, and the amount of moisture in the air, just to start with—a miscalculation on any of these can lead to a temperature forecast that can be off by as much as 10–20°F. Temperature doesn't normally change drastically one day to the next—but when it does, it's nice to be prepared. Here are some things to keep in mind. Note: Start with today's high and low temperature when forecasting for the future, and then adjust from there.

- Although the sun is directly overhead at noon, surface temperatures tend to reach their peak between 2 and 4 p.m. Not so coincidentally, that's when a lot of summertime thunderstorms occur.
- Calm winds and clear skies allow the warmth from the surface of the Earth to rise, and temperatures can drop rapidly during the night. (Windy clear nights are less cold.)
- Your high temperature can be lowered if you expect the arrival of a cold air mass, high wind speed, greater cloud cover, and higher humidity.
- You might increase your forecasted high temperature if you expect a warm front, low or nonexistent winds, less cloud cover, and lower humidity.

- Your best guess for a low temperature can be lowered if you expect a cold front, slow or nonexistent winds, less cloud cover, and less humidity.
- You might increase your best low temperature guess if you expect warm air mass, high winds, greater cloud cover, and more humidity.
- Your local geography has a greater impact on temperature, and this can vary from season to season. A lake in summertime can retain heat, but in winter reflect heat back to the atmosphere when frozen, keeping you slightly colder.
- With light and variable winds, heat increases at the surface but does not mix with the cooler air above (the opposite of what usually happens).
- Likewise, light wind at night does not mix the cool air at the surface with the warmer air above, keeping the temperature lower.
- A strong cold front will mean that the high temperature will occur before frontal passage, even if the front passes at, say, 10:00 a.m. Temperatures will tend to crease after 10 a.m.
- If you are relying on a thermometer to tell you the temperature, keep it in a protected area away from winds and out of direct sun.
- If you are forecasting the temperature for an outdoor event and there is a possibility the temperatures will be extreme (say, lower than 40°F or greater than 85°F), use either a wind chill chart or a heat index chart (both of which are measurements of what the outside temperature feels like on human skin) to make sure there are no ill effects that could affect health. (Both can also be found at http://www.srh.noaa.gov/ffc/html/metcalc.php.)
- Long-range temperature forecasts may give you trends, but the farther out you are, the more difficult it is to forecast.

NWS Windchill Chart

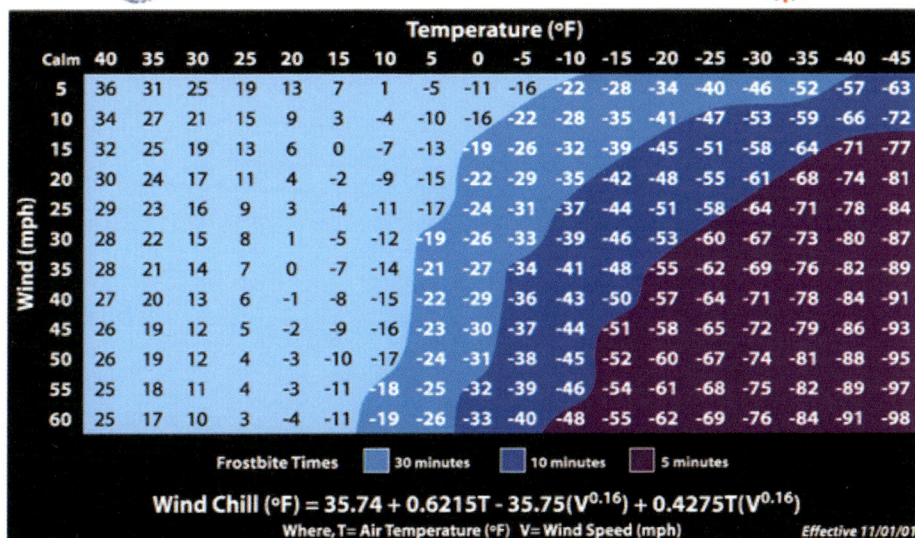

	Temperature (°F)																	
Calm	40	35	30	25	20	15	10	5	0	-5	-10	-15	-20	-25	-30	-35	-40	-45
5	36	31	25	19	13	7	1	-5	-11	-16	-22	-28	-34	-40	-46	-52	-57	-63
10	34	27	21	15	9	3	-4	-10	-16	-22	-28	-35	-41	-47	-53	-59	-66	-72
15	32	25	19	13	6	0	-7	-13	-19	-26	-32	-39	-45	-51	-58	-64	-71	-77
20	30	24	17	11	4	-2	-9	-15	-22	-29	-35	-42	-48	-55	-61	-68	-74	-81
25	29	23	16	9	3	-4	-11	-17	-24	-31	-37	-44	-51	-58	-64	-71	-78	-84
30	28	22	15	8	1	-5	-12	-19	-26	-33	-39	-46	-53	-60	-67	-73	-80	-87
35	28	21	14	7	0	-7	-14	-21	-27	-34	-41	-48	-55	-62	-69	-76	-82	-89
40	27	20	13	6	-1	-8	-15	-22	-29	-36	-43	-50	-57	-64	-71	-78	-84	-91
45	26	19	12	5	-2	-9	-16	-23	-30	-37	-44	-51	-58	-65	-72	-79	-86	-93
50	26	19	12	4	-3	-10	-17	-24	-31	-38	-45	-52	-60	-67	-74	-81	-88	-95
55	25	18	11	4	-3	-11	-18	-25	-32	-39	-46	-54	-61	-68	-75	-82	-89	-97
60	25	17	10	3	-4	-11	-19	-26	-33	-40	-48	-55	-62	-69	-76	-84	-91	-98

Wind (mph) (vertical axis label)

Frostbite Times: ▢ 30 minutes ▢ 10 minutes ▢ 5 minutes

$$\text{Wind Chill (°F)} = 35.74 + 0.6215T - 35.75(V^{0.16}) + 0.4275T(V^{0.16})$$

Where, T= Air Temperature (°F) V= Wind Speed (mph)

Effective 11/01/01

NOAA's National Weather Service

Heat Index

Temperature (°F)

	80	82	84	86	88	90	92	94	96	98	100	102	104	106	108	110
40	80	81	83	85	88	91	94	97	101	105	109	114	119	124	130	136
45	80	82	84	87	89	93	96	100	104	109	114	119	124	130	137	
50	81	83	85	88	91	95	99	103	108	113	118	124	131	137		
55	81	84	86	89	93	97	101	106	112	117	124	130	137			
60	82	84	88	91	95	100	105	110	116	123	129	137				
65	82	85	89	93	98	103	108	114	121	128	136					
70	83	86	90	95	100	105	112	119	126	134						
75	84	88	92	97	103	109	116	124	132							
80	84	89	94	100	106	113	121	129								
85	85	90	96	102	110	117	126	135								
90	86	91	98	105	113	122	131									
95	86	93	100	108	117	127										
100	87	95	103	112	121	132										

Relative Humidity (%) (vertical axis label)

Likelihood of Heat Disorders with Prolonged Exposure or Strenuous Activity

▢ Caution ▢ Extreme Caution ▢ Danger ▢ Extreme Danger

To use the heat index, you need relative humidity—and humidity is one of the most difficult measurements for the amateur. The easiest way is simply to purchase an inexpensive hygrometer (see the chapter 12 on weather equipment).

8   SEVERE WEATHER

Severe thunderstorms, tornadoes, hurricanes, blizzards, ice storms, and flash floods are all dangerous. If you think any of these is likely, get to safety. Rely on experts. Do not take chances. Experts have professional resources that can help predict movement and strength of storms and water runoff. If they issue a watch or a warning, take it seriously.

That said, here is a brief description of each and some of the weather patterns to look for.

Severe Thunderstorms

Thunderstorms are local cumulonimbus storms accompanied by thunder and lightning, and sometimes with strong wind gusts, heavy rain, and/or hail. Summer thunderstorms are not uncommon, and by themselves do little damage. The attendant lightning, however, sparks fires, causes electrical failure, and occasionally kills. Severe thunderstorms have one or more of the following:

- Winds greater than 57 mph
- Hail greater than 3/4" (the size of a nickel)
- Funnel cloud or tornado

There are "air mass" thunderstorms—usually due to heating of the air, geography, and occasionally to the diurnal oscillations of the jet stream—and there are frontal thunderstorms, which are more likely to be severe or to form a line, often shown as lines of red (or in a red box) on weather maps. Supercell thunderstorms have a strong updraft, and a mesoscale convective complex is a cluster of storms that moves slowly through an area (often causing flooding). Severe thunderstorms have significant lightning and local flooding. Hail and macro/microbursts often occur as well.

How can you predict a severe thunderstorm? The presence of cumulonimbus clouds and afternoon heating produce air mass thunderstorms; if you have frontal passage as well, severe thunderstorms may be more likely.

Tornadoes

Professional weather forecasters use Doppler radar to predict tornadoes and look for the signature shapes of storms where tornado formation might occur. Some of the things you can look for take into consideration are:

- Time of year. Tornadoes tend to appear in the South during February through April. In the northern Midwest and northern Plains, June through August. These are not definite dates, however; tornadoes have appeared at all times and all over the globe. Tornadoes simply need low-level heat, moisture, and instability in the right combination over the right geography.
- Time of day. Tornadoes tend to appear between 2 p.m. and 7 p.m., which is also the time of peak heating. However, they can still happen at any time of day.
- Geography. Tornadoes have an easier time to develop over plains and flat land, but hilly and mountainous areas can't be ruled out.
- Clouds. Tornadoes tend to develop from cumulonimbus mammatus, a subtype of cumulonimbus with low, hanging bumps.
- Debris. You can't always see the funnel for all the debris that is flying around. Don't wait till you see a funnel cloud.
- Warning signs include green skies and hail (even when there is no rain). Of course, by the time you see the hail, it may be too late to move safely. (Note: The skies aren't actually green, but the red of the sky near late afternoon illuminates blue water droplets, making them appear green. Tall cumulonimbus that do not form tornadoes also display this color.)

If you are out and feel a tornado is imminent, get to shelter immediately.

- *Water spouts are the same as a tornado, only over water. Most of the same rules apply.*

- *Not all tornadoes are counterclockwise—some (about 20%) rotate clockwise.*

- *Just like in a hurricane, extreme low pressure marks the center of the tornado. Atmospheric pressure can fall as rapidly as 34 mb in 15 minutes.*

- *The fastest wind speed in a tornado recorded so far is 318 mph. Don't take chances.*

Hurricanes

It would almost be impossible to miss the hurricane news and warnings that are prevalent these days, at least in the United States. Hurricanes develop in the Intercontinental Tropical Zone and their paths, though not predictable, generally follow an expected pattern once they reach cooler waters or land. Forecasters have gotten very good at predicting a hurricane's path. Pay attention to hurricane warnings.

Hurricanes begin as a tropical disturbance, then strengthen into a tropical depression. When the winds reach 39 mph (34 knots), it becomes a tropical storm and is given a name. When winds reach 64 knots (74 mph), it is a hurricane.

The winds in a hurricane move counterclockwise in the Northern Hemisphere, and can be up to 300 miles away from the eye. Hurricane season is from June 1 to November 30. Warnings are issued at least 36 hours in advance to allow preparation and evacuation.

Pressure inside a strong hurricane (measured in millibars) could be 920 mb or lower, while a weak hurricane might be greater than 980 mb. Since air is trying to move to an area of lower pressure, and the lower the pressure, the faster the air is moving, this leads to winds greater than 155 mph in a strong hurricane.

Blizzards

Like hurricanes, blizzards are often easy to predict (especially now with "ensemble" forecasts) though the degree of snow and hazard that they produce is not as easy to forecast. Blizzards are defined by any three of the following:

- Winds of over 35 mph
- Visibility less than a quarter mile
- Duration greater than 3 hours
- Temperature below 20°F

Like thunderstorms, blizzards require instability and moisture in the air. They are identified by areas of low pressure, wind flow patterns, temperature, and dewpoint (a measure of the moisture in air).

The aftereffects of a blizzard are sometimes as severe as the blizzard itself. The snow melts and then freezes again, sometimes forming enough ice to cause trees to fall and electricity (and heat) to fail. Freezing rain is treacherous.

How to predict snowfall totals is extremely difficult, and very easy to verify. If a forecaster expects 4" of rain and we get 6", very few people would notice the difference in rainfall. However, if 6" of snow is forecasted and we get 18", all it takes is a look out the window to see the difference—yet the difference in water content is less for the snow than the rain. (As a general rule, 1" of rain equals 10" of snow.)

Ice Storms

Ice storms are known for the havoc they wreak with regard to electricity and travel, both during and after the storm. They are hard to predict because the air through which the water droplets fall is not just freezing but is "supercooled"—causing anything the water molecules fall on to become coated with ice. The ice slowly builds up as more and more supercooled drops land in the same place. There are some theories that ice storms will begin to occur more frequently as global warming becomes more prevalent because of the likelihood of freezing rain instead of snow. Forecasters are learning a lot more about ice storms, and each winter their forecasts become more accurate.

Global warming—climate change—whatever words you want to use—in this respect I won't use weasel words. Our planet is getting warming, for better or for worse, and it's the only planet (so far) we've got. We need to take care of it.

Something like 97 percent of scientists agree that global warming is here and that we humans are creating problems—or, at the very least, exacerbating problems—that are going to make living here very difficult indeed. Do what you can to lessen the effects. One of the major problems scientists foresee is that weather extremes will be more pronounced—drier areas will become drier, wetter area will become wetter. Hurricanes (and, in the eastern Pacific, their equivalent, typhoons) and tornadoes will become stronger. But that's just the beginning.

There is no doubt that our planet is getting hotter, and heat acts as a trigger for all sorts of weather problems.

For your purposes, any storm in winter has the likelihood of becoming an ice storm. They are most likely to occur with cumulus clouds and freezing temperatures in the clouds (though not necessarily on the surface). A new index from NOAA has been developed (which won't help forecast a storm, but helps prepare residents):

Ice Index Parameters			
Index	Radial Ice Accumulation (Inches)	Wind (mph)	Damage and Impact Descriptions
1	0.10 – 0.25	15 – 25	Some local utility interruptions possible...typically lasting a few hours.
	0.25 – 0.50	> 15	
2	0.10 – 0.25	25 – 35	Scattered utility interruptions possible...typically lasting less than 12 hours.
	0.25 – 0.50	15 – 25	
	0.50 – 0.75	< 15	
3	0.10 – 0.25	≥45	Numerous utility interruptions possible...lasting up to 5 days. Damage to some main feeder lines possible.
	0.25 – 0.50	25 – 35	
	0.50 – 0.75	15 – 25	
	0.75 – 1.00	< 15	
4	0.25 – 0.50	≥ 35	Prolonged and widespread utility interruptions possible. Damage to many main feeder lines possible. Utility outages lasting up to 10 days possible.
	0.50 – 0.75	25 – 35	
	0.75 – 1.00	15 – 25	
	1.00 – 1.50	< 15	
5	0.50 – 0.75	≥ 35	Catastrophic damage to exposed utility systems possible. Outages lasting several weeks possible in some areas.
	0.75 – 1.00	≥ 25	
	1.00 – 1.50	≥ 15	
	> 1.50	Any	

The Damage and Impact Descriptions are based upon: (1) researched weather parameters and utility impacts and (2) the combination of **forecast** parameters including radial ice accumulation, wind and temperatures.

Flash Floods

A flash flood can have any number of causes , but for our purposes here, it is 2" or more of rain in a 12-hour period, although this can vary by geographic location. Mountainous areas and flat plains are particularly vulnerable, as are low-lying areas of cities. Snowmelt can also cause significant flash flooding. By definition, flash floods occur without warning. As little as 2' of water can sweep away automobiles. Be alert.

Flash floods can occur during a single storm, and most often occur when the ground is already saturated. Water no longer drains and instead rushes across the surface, sweeping everything in its path. One of the biggest problems with flash flooding is our expanding development in low-lying areas.

Storms that are likely to cause flash flooding are severe thunderstorms and intense storms of long duration (i.e., along a front, moving parallel to the front). By studying historical patterns of storms before they reach you, you can identify potential sources of flash floods.

9 GEOGRAPHY

No matter where you live, the geography around you affects the weather. Here are some ways that your surroundings can change the amount of rain/snow you get, the winds, the temperature, and so on.

Remember the 200-mile upstream guideline. If the wind is blowing from the southwest, look 200 miles to your southwest and, with exceptions, their weather today is your weather tomorrow. But if they're on the edge of a desert and you're on a mountaintop, that's a different story.

Coastal Areas

Obviously, moisture is readily available for afternoon thunderstorms during the summer. It doesn't take much to trigger one. This is also true for areas next to large bodies of water, especially those on the eastern shore of the body of water (since most weather in the Northern Hemisphere has a westerly component) most of the time.

Water is slower than land to warm up or to cool down—that's why you often see fog develop over smaller bodies of water, especially early in the morning.

Sea breezes occur on warm afternoons as the land warms up. Warm air rises from the land, causing the cooler air from the water to rush inland to replace it. In tropical areas, however, the opposite can happen in the evening. Land cools down quickly as soon as the sun goes down just as the warmer air over the ocean is rising—and the air rushes out to sea.

The abundance of warm water also contributes to the strengthening of storms, including thunderstorms, hurricanes, and blizzards. As warmer water moves northward, it feeds the storm, helping to continue the duration of the storm. There can be as much as a 10°F difference between the two temperatures of land and water. Once cooler waters intrude, the storm begins to subside. The transition zone between the cooler water of the ocean and the warmer Gulf Stream water acts like a front, causing turbulence and winds.

Pacific Coast weather is different than East or Gulf Coast weather, in large part because storms that appear on the West Coast have formed hundreds of miles away out over the Pacific (where we don't have radar) and have had time to build. The sheer variety of geographic formations in the West also increases the unpredictability of weather.

Mountains

Other than the greater chance of snow at high elevations, there are significant and lesser known weather differences for residents of mountainous areas. People who live on the windward and leeward sides of mountains have different experiences with storm systems. On the windward side (generally the westward side), drier air moves up along the mountain, cooling as it goes. Because warmer air can hold more water vapor, more precipitation falls out of the clouds as they rise and are less able to hang onto the moisture. The weather on the windward side is more like a moist, coastal area. Once that cold air reaches the top, it sinks on the leeward side, pushing out the warm air ahead of it. The air is significantly drier, acting more like a desert. This area has a dry climate and is known as a "rain shadow." (Palm Springs receives an average of 5.8 inches of rain; its windward companion, Beaumount, not even 30 miles away, has an average rainfall of almost 20 inches.)

Weather events in a rain shadow tend to be more severe because of the extreme differences between windward and leeward sides, and weather forecasting is particularly difficult. The variances in elevation just to the west create wind patterns and minor turbulence that vary by minute wind direction changes. Climatology studies (those of long-term duration) may be of more help than trying to forecast the weather by current conditions, though these must take the warmer global conditions into account (and hence even high amounts of water vapor).

Desert

Dry, right? Of course. Even though the sun might shine for 360 days a year, there is little to distinguish the storms that do cause rain from those that don't. To all appearances, the storms look the same—form in the same areas, have the same westerly component, and so on. It's easy to forecast a storm when none appears, and vice versa. The consequences of not predicting a storm, however, can be severe; sand does not soak up water quickly, and the resultant flash flooding can cause serious damage.

Sandstorms are violent whirlwinds of sand and dust. Visibility can be less than a quarter mile. In the flat and elevated areas of a desert, you can generally see the weather approaching; in lower elevations, however, storms can appear within very little warning.

Desert weather forecasters have a particularly hard time, because everyone remembers the times you got it wrong (maybe twice a year) from the times you got it right (360 days of sunshine plus maybe 3 days of rain…).

Urban Areas

Weather effects tend to be magnified over short periods in urban areas, possibly due to the heat island effect. Compared to the surrounding rural areas, temperatures are 4–10°F warmer in the city. Think of all the warm air over the vast areas of asphalt—streets, parking lots, and so on. The warm air rises, causing more instability when a storm passes through. At the same time, warmer air near the surface holds more moisture. These effects can build and intensify a storm. And because cold air rushes toward warmer air, all the surrounding winds are drawn to the city.

At the same time, wind whips around taller buildings, creating channels of even stronger wind. This helps create instability as well. Further downstream, past the buildings, the channels of wind merge, the warmer air rises, causing more clouds, and areas downstream see greater instability. In short, cities are unpredictable weather magnets.

With warmer temperatures and less area of shade to block the sun, hot air is trapped and heat waves are more intensified.

Air pollution (from automobiles, industry, or any other source) forms condensation nuclei, and the water vapor can develop around those minuscule particles.

Great Plains

With nothing else to do for hundreds of miles, weather systems feed on themselves. Polar air masses move south from Canada into the Midwest at the same time warmer air masses move from the west and south. These collisions lead to more instability, which naturally leads to more—and more violent—storms in all seasons. Wind speeds tend to be very high—and tornadoes are prevalent.

Did I Miss an Area?

Write me at rainorshine@swiftpassage.com and let me know.

10 WEATHER PROVERBS

There are many, many weather proverbs. Some of the more accurate common ones are listed here. Most are not accurate over the long term, many are purely circumstantial, and some are not yet proven. If you have another one that seems to make sense, please send it to: rainorshine@swiftpassage.com. And note that most of these apply to the continental United States only. Something valid here—or valid in a specific location—might not make sense elsewhere.

"Red [or pink] sky at night, sailor's delight; Red sky at morning, sailors take warning."

Everyone's heard this one, it actually has a basis in science, and it even appears in the Bible and Shakespeare in vaguely similar words. Weather tends to move from west to east. Red skies in the morning are caused by the sun's reflection off the minute particles in the air. If they are in the west, those particles—which might later be the condensation nuclei of a storm— are moving your way. If you see them in the evening, those reflected particles will have already passed you by.

But this only works if you think of "warning" and "storm clouds/rain/high winds" as the same thing. Sailors also have problems with a lack of wind— after all, it's hard to sail in no wind at all—in which case, this part of it doesn't work out at all. Substitute "shepherd" for "sailor" (an older version of the same proverb) and it's generally true.

"With the wind at your back, stick your left arm out. Your fingers will point in the direction of the storm."

With modification, this one is somewhat true. Remember that winds travel in a cyclonic (counterclockwise) direction in a low-pressure (read: stormy weather) system. If you stand with the wind at your back and extend your left arm, you will be pointing toward a low-pressure system. Whether that low-pressure system will affect your weather in any meaningful way is not so certain—the wet weather might be far enough away that you won't get much, or it may already have passed you by if it's a relatively dry front, but with a gust front ahead of it.

"If the bubbles move to the rim, a storm is sure to brim. If they sink in the middle, the day'll be fit as a fiddle."

I had to try this one to be sure it worked, and surprisingly, it does. (Coffee advertisers should pick this up, don't you think?) Pour yourself a cup of coffee or tea and watch the bubbles. If the barometric pressure is fairly high, the surface of the coffee is concave and the coffee clings to the edges, letting the bubbles float to the center. When pressure is low, the surface is convex, and the center rises—pushing the bubbles to the outside. But even here, if the low-pressure system is a weak one and isn't bringing with it any precipitation, you won't see a storm. (Note: If you put the cream in first, I'm less sure of its forecasting abilities...more experimentation is needed.)

"Mackerel scales and mares' tails make lofty ships to carry low sails."

Mackerel scales and mares' tails refer to the cirrocumulus clouds, which often precede a warm front. As winds shift from an easterly component (usually northeast or east) to a westerly component (from the west or southwest), they often bring with them rain within 24–36 hours.

"Clear moon, frost soon."

Clear nights allow the heat buildup in the Earth to rise and colder air to move in. Obviously this one isn't true in summer in the South when temperatures remain about 70°F (a little frost would be more than welcome, though pretty much impossible), but in the fall or winter, a clear night can portend cooler than expected nighttime temperatures.

"Halo around the sun or moon, rain or snow will come very soon."

On mostly clear nights where the moon is easily visible, a halo appears around it. This is the reflection of the sunlight reflected by the moon onto the ice crystals of thin cirrus, indicating the presence of water vapor in the air—which is often a precursor to wetter weather. In general, the brighter the halo, the more ice crystals present. Sometimes they say you can even see the halo around the sun (though you should never look directly at the sun, so how do they know?).

There are many more proverbs, and I enjoy collecting them. Sometimes the most surprising ones turn out to be true. If you have one that tends to be true, send it to me at rainorshine@swiftpassage.com and be sure to let me know where you are.

11 WEATHER EQUIPMENT

You really don't need much:

- Paper
- Pen
- Outdoor thermometer
- Barometer
- Anemometer or some way of measuring wind speed and direction (either an instrument or a window)
  And nice to have but not really necessary for forecasting:
- Hygrometer (for relative humidity, if needed, for the heat index; if you're in Alaska, don't worry about this one!)
- Rain gauge (handy, though it's a measurement of what's already happened)

That's it. An Internet connection helps because you can check upstream weather more precisely (but the television does a pretty good job). Cloud books are available from your local library or for purchase in brick-and-mortar stores or online. I won't make recommendations as to brand for the other pieces, and you can find weather stations or individual pieces of equipment online, at flea markets, and just about everywhere and in every price range.

Barometer

Figure 12.1. Aneroid barometer

Your barometer will need to be calibrated to your location. For weather forecasting purposes, barometers are calibrated to show what they would be if they were actually at sea level so that all barometers can be compared. If you are in a valley and the nearest barometer is on a mountaintop, your pressure would tend to be much higher than the other, even as low-pressure systems move overhead. How would you know when a low-pressure or high-pressure system reached you, if their atmospheric pressure is always lower than yours? Imagine the confusion if every barometer reported its own pressure. Someone somewhere would have to convert all the readings to something standard (allowing for elevation) before the front or the system could be found. Trends between any two locations would be hard to pinpoint. It's much easier to calibrate all barometers to sea level pressure, and they can then be compared on an equal basis.

Even though trends will still be reliable without calibration at each individual location, accurate current readings are important for comparison with other locations. Here's how to do that for aneroid ("without air") barometers. Note: Aneroid barometers tend to be less accurate than mercury barometers, but the latter is increasingly hard to find and does have potential hazards. Most barometers today are aneroid (analog) or electric (digital).

- Go to weather.gov, find your city, and get the most recent reading (OR) if you are in a remote area, find a local airfield and check with a pilot or airport personnel (OR) as a last resort and recognizing that this will not give a very accurate reading, get a barometric reading from the closest city and subtract 1 mbar for every 10 meters increase in elevation (or add 1 mbar for every 10 m decrease in elevation).

- Adjust your barometer by means of whatever adjustment mechanism your barometer has (some have a set screw on the back that must be adjusted, others a small dial like a clock) to the reported reading.
- Recalibrate your barometer about once a year.

Figure 12.2. Digital barometer. Note that the trend is for the past 24 hours.

Setting a digital barometer varies, depending on manufacturer. Refer to your instruction booklet. In general, some require you to input your elevation and the barometer takes care of the adjustment. Others require digital input to begin with and adjust from there.

Note that temperature changes, especially abrupt ones, can affect your pressure readings and will require another calibration. Try to avoid such changes, but if they do happen, simply recalibrate your barometer.

Anemometer

An anemometer is an instrument that measures wind speed and direction. Although this instrument isn't strictly necessary, it is significantly more accurate than standing in a field away from anything (trees, buildings, and so forth) that would change the wind pattern or speed. Most anemometers have two components: a reading device and a measuring device. The measuring device is located in a place away from obstructions—on a rooftop or some other location where the surrounding area does not affect the wind flow, like a weather vane or a windsock. It usually consists of several small cups that catch the wind, and the results are read on the much more conveniently located reading device.

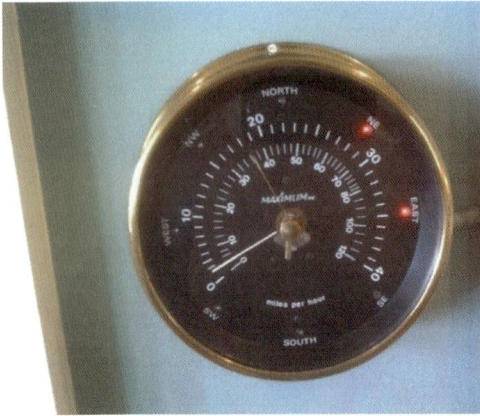

Figure 12.3. Analog anemometer display.

The example here shows the wind blowing at about 2 mph from the east-northeast. Does this easterly wind indicate that wet weather might be coming? Not really. The wind moments later was showing south-southwest. Anything below about 6 mph tends to be variable. Also, the pressure shown in the above photo indicates 30.00—a relatively high, steady pressure. If the wind had been 10 mph, then I would consider forecasting a change in the weather.

Like barometers, they are both digital and analog, but unlike barometers, no calibration is necessary. Refer to the manufacturer's instructions for initial setup.

Hygrometer

A hygrometer measures the amount of moisture in the air. Hygrometers rely on calibration of temperature, pressure, changes in electrical resistance, and other methods to calculate humidity. There are various types of hygrometers and sensors, and you should refer to your manufacturer's instructions to keep your instrument in peak condition. These tend to be used for heat index calculations, as well as dewpoint and other moisture measurements.

Absolute humidity is the amount of water vapor in the air (per unit of volume; e.g., 25 grams per cubic meter), while relative humidity is the amount of water vapor expressed as a percent of the amount of water vapor that could be in the air at a particular temperature (e.g., 61%). Relative humidity is what we use for heat index calculations and to determine the amount of moisture in the air masses. (We could use absolute humidity—but only if we knew exactly how much volume the oncoming air mass took up.

Hygrometers should arrive already calibrated and should include instructions for recalibration. They are simple to read—just read the number at the dial for analog, or the number off the screen for digital. No corrections are necessary.

Sling psychrometers do the same thing by calculating wet bulb and dry bulb temperatures, but are harder to use.

Rain Gauge

These are usually simple tubes with easy-to-read measurements on the outside. Rain gauges should be emptied after each rainfall.

12 WEATHER MAPS

Weather maps are not difficult to read when you take the bits and pieces apart. Here's a typical U.S. surface map:

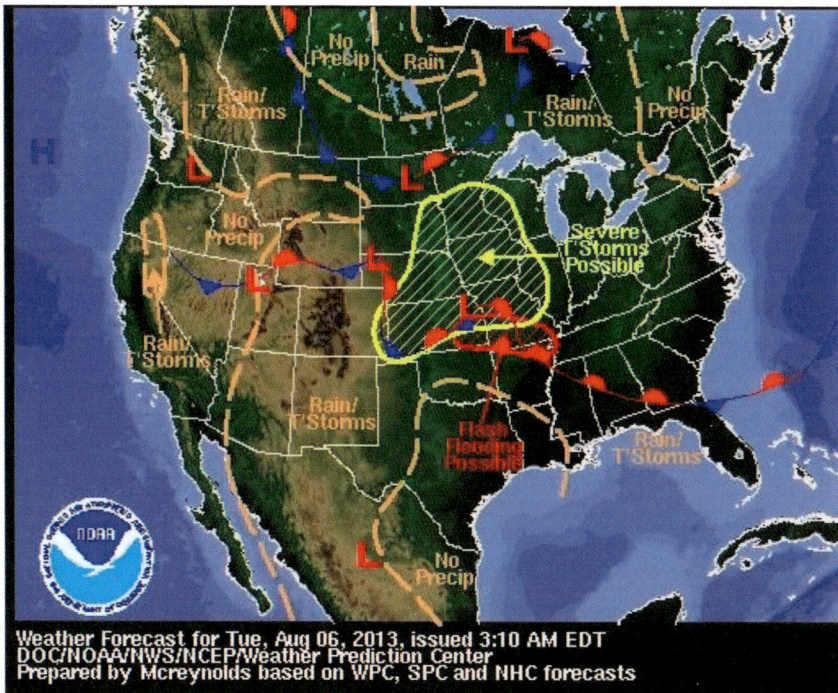

Figure 13.1. Surface weather forecast.

Cold fronts are lines of blue triangles, warm fronts are lines of red curves, stationary fronts are mixed red and blue, and occluded fronts (none on this map) are purple. The blue H is a high-pressure systems, the red Ls are low-pressure systems. There's possibly a high-pressure system is also in the northeast, but it isn't marked and may not be significant. Technically, there's also high pressure between any two lows (e.g., in the map above, South Dakota is probably experiencing high(er) pressure), but in many cases the areas are small and the weather is not noticeably different.

Here are some things to notice:

- The areas of precipitation follow the cold fronts and are pushed ahead of the warm fronts.
- Where the low-pressure systems are converging is a strong possibility of severe weather. Note how all those lows (moving west to east) are pushing out the former high-pressure system that dominated the east and is probably now just offshore New Jersey.
- Flash flooding looks possible in the Ozarks, probably especially in valleys and along rivers.

A forecasted precipitation map for the next twelve hours looks like this. The outer (brown) lines indicate a trace (or less than .1" of rain), the next inward (red) lines indicate .25", the next inward (green) lines indicate either .5" or .6" as marked, and the small centers (blue) lines indicate either 1" or 2" of rain forecasted. Remember that 2" of rain in less than 24 hours means an increased possibility of flash floods, which is marked above.

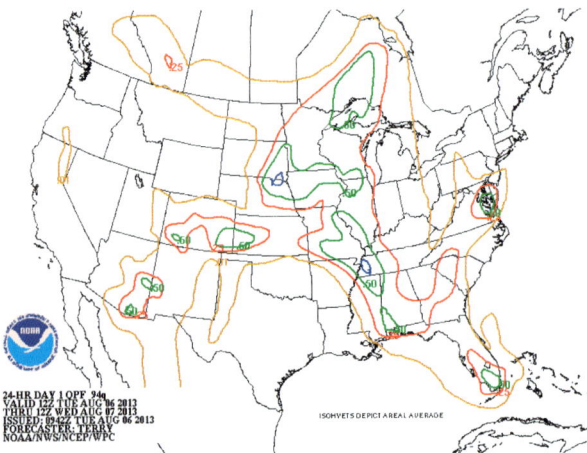

Weather maps will often also

Figure 13.2. Rainfall prediction. (Drawing courtesy of NOAA.)

show isobars—lines of equal atmospheric pressure. These are normally indicated in millibars.

Winds are shown as small barbed arrows, showing the direction the wind is from. (In this chart, there is no significant wind, and the directions are varying all over. Notice how off the coast of New Jersey the winds are going in a clockwise direction, indicating that possible high-pressure system.)

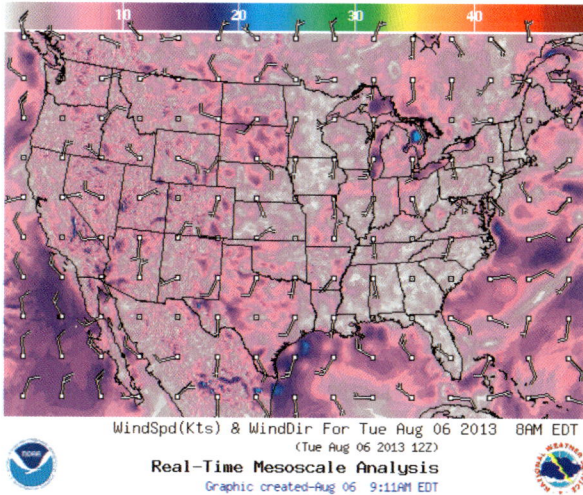

WindSpd(Kts) & WindDir For Tue Aug 06 2013 8AM EDT
(Tue Aug 06 2013 12Z)
Real-Time Mesoscale Analysis
Graphic created-Aug 06 9:11AM EDT

Figure 13.3. Wind report (courtesy of NOAA).

Maps will have legends, such as this one, and many are digitally enhanced. You can create your own surface maps from weather reports upstream using conventional codes or creating your own.

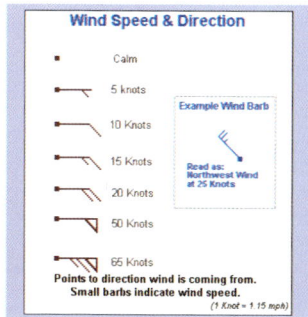

Figure 13.4. Legend for wind reports showing speed and direction.

All of these maps, charts, and legends are from the National Oceanic and Atmospheric Administration (NOAA) at www.noaa.gov. The site is an excellent resource for all things weather related, including education, current and forecast conditions, weather warnings, and alerts.

I find that their weather forecasts are generally accurate (once I put my own "spin" on them, considering where I am at any particular time). They are unbeatable for warnings, alerts, and serious weather conditions. They can give you the upstream weather if you have no other local resources. Use them, learn from them, and support them.

13   PUTTING IT ALL TOGETHER—MAKING YOUR OWN FORECASTS

It's time to create your own forecasts. In the following pages, I've created worksheets geared to 5-day forecasts, 3-day forecasts, 2-day forecasts, and next day forecasts. Practice with them before your "big event" (if you have one coming up). Learn your local weather quirks. Maybe the 200-mile upstream rule should be 180 miles for you, maybe 250. Maybe that highway that runs north–south just six miles away acts like a heat barrier and keeps the rain from reaching you most of the time. Maybe you live just close enough to the leeward side of a mountain that you get more precipitation than your neighbors a mile away.

Once you have your local winds, use Google Maps or any other source to locate your upstream stations. Unless you're lucky enough to live in the center of the country, chances are you'll be offshore or in Mexico (where weather observations are less easy to obtain) for your 800-mile forecast. Don't worry if you can't find one. I like Google Maps because it has an easy-to-use scale in the lower left corner, but any source with a scale will do.

Then use www.noaa.gov to find the weather at your upstream stations. By plugging in the city name and state, you'll get an easy-to-understand forecast, but for trends click on the link for "3-day history."

Following the blank worksheets is one set I completed, copied over from my scribbled pages to make the handwriting more legible and expanded upon to show my thinking. (If you are reading this on an electronic reading device and want a copy you can duplicate, email me at rainorshine@swiftpassage.com and I'll forward you a set that you can print and copy.)

Some tips for completing these worksheets:

- If you have a fixed time in mind for your 5-day forecast—say, a wedding at 4:00 p.m. on Saturday—do your forecasts during the week leading up to it at around the same time of day. For example, you'll want to know whether afternoon thundershowers in the summer are likely—but if you do your forecasts at 11:00 a.m. when the weather is clear, you might not be aware of the trend of afternoon storms.
- When checking the weather at your upstream stations, glance over the day's reports and see if there's something that triggers your interest: a 10-degree difference between one hour and the next, indicating frontal passage, or maybe a strong wind shift. The 3-day history will also give you a pressure trend.
- I used weather abbreviations on my worksheets, and you can find a list of them at Wikipedia under "Station Model." However, feel free to make up your own abbreviations as needed.

Have fun, and enjoy the weather, whatever it is.

5-Day Weather Forecast

Today's Date: _____   Forecast Date: _____

Current weather here:

- Cloud cover:
- Pressure:
- Precipitation (if any):
- Wind direction and speed:
- Temperature (high/low):

What is the weather 200 miles upstream? (Where is this? _____)

- Cloud cover:
- Pressure:
- Precipitation (if any):
- Wind direction and speed:
- Temperature (high/low):

> Go to www.noaa.gov or other reliable source and insert data for each of these locations.

What is the weather 400 miles upstream? (Where is this? _____)

- Cloud cover:
- Pressure:
- Precipitation (if any):
- Wind direction and speed:

What is the weather 600 miles upstream? (Where is this? _____)

- Cloud cover:
- Pressure:
- Precipitation (if any):
- Wind direction and speed:

What is the weather 800 miles upstream? (Where is this? _____)

- Cloud cover:
- Pressure:
- Precipitation (if any):

- Wind direction and speed:
- Are there any severe weather warnings or alerts anywhere along the path? If so, what?

What local conditions could affect your forecast?

Based on your 800-mile forecast and your local conditions, what do you expect on your forecast date?

3-Day Weather Forecast

Today's Date: _____   Forecast Date:
_____

Current weather here:

- Cloud cover:
- Pressure:
- Precipitation (if any):
- Temperature (high/low):
- Wind direction and speed:

What is the weather 200 miles upstream? (Where is this?
_____)

- Cloud cover:
- Pressure:
- Precipitation (if any):
- Temperature (high/low):
- Wind direction and speed:

What is the weather 400 miles upstream? (Where is this?
_____)

- Cloud cover:
- Pressure:
- Precipitation (if any):
- Wind direction and speed:

What is the weather 600 miles upstream? (Where is this? _____)

- Cloud cover:
- Pressure:
- Precipitation (if any):
- Wind direction and speed:
- Are there any severe weather warnings or alerts anywhere along the path? If so, what?

Obtain the data from the same source as you used for the 5-day forecast.

Is the previous 600 - mile forecast in line with your current 200-mile observation? If not, have the winds changed (in which case all your upstream locations should have changed)? Is the weather moving more slowly or more quickly?

What has changed from your initial worksheet? And more importantly, why?

67

Review the Wind/Barometer Table. What does it tell you will happen? Does this agree or disagree with the current 200-mile conditions? If it disagrees, are the differences significant?

Review the Weather Changes Due to Frontal Passage table. Does it apply? What does it tell you will happen? Does this agree or disagree with current 200-mile conditions? If it disagrees, are the differences significant?

Review the Forecasting by Clouds table. What does it tell you will happen? Does this agree or disagree with the current 200-mile conditions? If it disagrees, are the differences significant?

Based on your 600-mile forecast and your local conditions, what do you expect on your forecast date?

2-Day Weather Forecast

Today's Date: _____   Forecast Date:
_____

Current weather here:

- Cloud cover:
- Pressure:
- Precipitation (if any):
- Temperature (high/low):
- Wind direction and speed:

> Obtain the data from the same source as you used for the 3-day forecast.
>
> Is yesterday's 400-mile forecast in line with today's 200-mile observation? If not, have the winds changed (in which case all your upstream locations should have changed)? Is the weather moving more slowly or more quickly?

What is the weather 200 miles upstream? (Where is this?
_____)

- Cloud cover:
- Pressure:
- Precipitation (if any):
- Temperature (high/low):
- Wind direction and speed:

What is the weather 400 miles upstream? (Where is this? _____)

- Cloud cover:
- Pressure:
- Precipitation (if any):
- Temperature (high/low):
- Wind direction and speed:
- Are there any severe weather warnings or alerts anywhere along the path?

Your forecast is just over 48 hours away. What has changed from yesterday's forecast?

Revisit the three tables—Wind/Barometer table, Weather Changes Due to Frontal Passage, and Forecasting by Clouds. Has anything changed from the forecast you prepared yesterday? (If anything has changed significantly, what was it and how did you miss it?)

What is your revised forecast for two days away?

Next-Day Weather Forecast

Today's Date: _____   Tomorrow's Date: _____

Current weather here:

- Cloud cover:
- Pressure:
- Precipitation (if any):
- Temperature (high/low):
- Wind direction and speed:

What is the weather 200 miles upstream? (Where is this? _____)

- Cloud cover:
- Pressure:
- Precipitation (if any):
- Temperature (high/low):
- Wind direction and speed:
- Are there any severe weather warnings or alerts anywhere along the path? If so, what?

What does the Wind/Barometer table tell you will happen tomorrow?

What does the Weather Changes Due to Frontal Passage table tell you will happen tomorrow?

What does the Forecast by Clouds tell you?

What does the afternoon temperature tell you? (i.e., are storms going to build up because of heat?)

What do the overnight temperatures and clouds (or lack thereof) tell you?

How does this agree or disagree with your 200-mile upstream observation?

If you are forecasting a frontal passage between now and tomorrow, how fast is the front moving? Is the rain or snow ahead (typical of warm front) of the front or behind it (typical of cold front)?

71

Go outside and listen to your instincts. Are you getting any conflicting messages? (e.g., is your forecast for clear weather but it smells/feels like rain?)

So, what is your forecast for tomorrow? Write it down here.

5-Day Weather Forecast

Today's Date: AUG 13   11 am   Forecast Date: 4:00 pm SUNDAY, AUG 18

Current weather here:

- Cloud cover: CLR
- Pressure: 29.97↑
- Precipitation (if any): —
- Wind direction and speed: Vrbl 7
- Temperature: 87

TS in p.m.

What is the weather 200 miles upstream? (Where is this? ALBANY, GA   )

BROKEN↓   ↙OVERCAST
- Cloud cover: 020 ① 038 ⊕
- Pressure: 30.04↓
- Precipitation (if any): —
- Wind direction and speed: W7
- Temperature:

NOTE PRESSURE
TRENDS — ↑↓↑
= POSSIBLE FRONT
NEAR ALBANY
(POINT OF LOW
PRESSURE)

What is the weather 400 miles upstream? (Where is this? MOBILE, AL   )

SCATTERED
- Cloud cover: 025 ⦶ 040 ①
- Pressure: 30.05↑
- Precipitation (if any): —
- Wind direction and speed: NW 5
- Temperature: 89

What is the weather 600 miles upstream? (Where is this? AUSTIN, TX   )

- Cloud cover: CLR
- Pressure: 30.05 —
- Precipitation (if any): —
- Wind direction and speed: SW 7

What is the weather 800 miles upstream? (Where is this? MEXICO   )

- Cloud cover:
- Pressure: ✗
- Precipitation (if any):
- Wind direction and speed:
- Are there any severe weather warnings or alerts anywhere along the path? If so, what?
     NONE

What local conditions would affect your forecast?

Based on your 800-mile forecast and your local conditions, what do you expect on your forecast date?
     possible pm TS

3-Day Weather Forecast

Today's Date: 8-15 12 pm   Forecast Date: 4:00 SUNDAY AUG 18

Current weather here:

- Cloud cover: 009 ⊕ AS/ST
- Pressure: 30.01 ↓
- Precipitation (if any): —
- Wind direction and speed: NE 13
- Temperature: 74 DEWPOINT 74   → Ts in p.m.
  = 100% humidity ⟶

What is the weather 200 miles upstream? (Where is this? WILMINGTON, NC)

- Cloud cover: 007 ① 012 ① 028 ⊕   NOTE CLOUDS COMING
- Pressure: 30.11 ↑   OUR WAY, PLUS
- Precipitation (if any): —   PRESSURE
- Wind direction and speed: N 13   DIFFERENCE.
- Temperature:

What is the weather 400 miles upstream? (Where is this? ELIZABETH CITY, NC)

- Cloud cover: 046 ⊕
- Pressure: 30.13 –
- Precipitation (if any): —
- Wind direction and speed: N 8
- Temperature:

What is the weather 600 miles upstream? (Where is this? SALISBURY, MD)

- Cloud cover: —   NOTE WINDS SHIFTING
- Pressure: 30.17   FROM NE TO NW
- Precipitation (if any): —   ...?
- Wind direction and speed: NW 3
- Are there any severe weather warnings or alerts anywhere along the path? If so, what?

Review the Wind/Barometer Table. What does it tell you will happen? Does this agree or disagree with the current 200-mile conditions? If it disagrees, are the differences significant?

Review the Weather Changes Due to Frontal Passage table. Does it apply? What does it tell you will happen? Does this agree or disagree with current 200-mile conditions? If it disagrees, are the differences significant? WARM FRONT, RAIN LIKELY
— RAIN

Review the Forecasting by Clouds table. What does it tell you will happen? Does this agree or disagree with the current 200-mile conditions? If it disagrees, are the differences significant? Based on your 600-mile forecast and your local conditions, what do you expect on your forecast date?

rain unless we get stronger winds

74

2-Day Weather Forecast

Today's Date: 8-16 1pm   Forecast Date: 4:00 SUN., AUG 18

Current weather here:

- Cloud cover: 100 ⊕ AS
- Pressure: 29.94 ↓
- Precipitation (if any): none now, but rain/TS yesterday pm + during night
- Wind direction and speed: vrbl 5
- Temperature: 87

NOTE: WINDS ARE NOT STRONG ENOUGH TO MOVE RAIN OUT OF AREA

What is the weather 200 miles upstream? (Where is this? CHARLOTTE, NC)

- Cloud cover: 035 ① 050 ⊕ — more clouds probably mean not much change
- Pressure: 30.14 —
- Precipitation (if any): —
- Wind direction and speed: NE 9
- Temperature: 71

What is the weather 400 miles upstream? (Where is this? PARKERSBURG, WV)

- Cloud cover: 045 ①
- Pressure: 30.14
- Precipitation (if any): —
- Wind direction and speed: —
- Temperature:
- Are there any severe weather warnings or alerts anywhere along the path?
  flash floods

Your forecast is just over 48 hours away. What has changed from yesterday's forecast? NOT MUCH, BUT IF WINDS PICK UP, THE RAIN MIGHT CLEAR OUT

Revisit the three tables--Wind/Barometer table, Weather Changes Due to Frontal Passage, and Forecasting by Clouds. Has anything changed from the forecast you prepared yesterday? (If anything has changed significantly, what was it and how did you miss it?) more rain

What is your revised forecast for two days away?
RAIN ESP. IN P.M.

Next-Day Weather Forecast

Today's Date: 8-17 3pm   Tomorrow's Date: 4:00 SUN, AUG 18

Current weather here:

- Cloud cover: O10 ⊕ CB/TS
- Pressure: 30.10↓
- Precipitation (if any): rain, heavy at times
- Wind direction and speed: SW 2
- Temperature: 85

What is the weather 200 miles upstream? (Where is this? ALBANY, GA) ← SAME AS 5-DAY

- Cloud cover: O14 ⑪, 110 ⊕
- Pressure: 30.08↓
- Precipitation (if any): Lt rain
- Wind direction and speed: SE 7
- Temperature:
- Are there any severe weather warnings or alerts anywhere along the path? If so, what?
  flash floods till 11pm

What does the Wind/Barometer table tell you will happen tomorrow?
fair, not much change

What does the Weather Changes Due to Frontal Passage table tell you will happen tomorrow?
going back + forth — occluded front

What does the Forecast by Clouds tell you?
increasing As ⟹ more rain, CB ⟹ TS

What does the afternoon temperature tell you? (i.e., are storms going to build up because of heat?).
storms, but not due to heat (won't pass quickly)

What do the overnight temperatures and clouds (or lack thereof) tell you?
warmer

How does this agree or disagree with your 200-mile upstream forecast?
more rain

If you are forecasting a frontal passage between now and tomorrow, how fast is the front moving? Is the rain or snow ahead (typical of warm front) of the front or behind it (typical of cold front)?
front overhead — back + forth

Go outside and listen to your instincts. Are you getting any conflicting messages? (e.g., is your forecast for clear weather but it smells/feels like rain?)
raining, clearer toward west

So, what is your forecast for tomorrow? Write it down here.

Rain likely in p.m. unless strong winds in a.m. push front away - but strong winds not likely.
SW wind moving westward.

76

14 CONCLUSION

That's it. It's much easier to forecast for a small area than a large one. After each of your forecasts, evaluate how you did so that you won't make that mistake again. Weather is unpredictable in many of its facets. If you got the "big ticket" items right, pat yourself on the back. If you blew it (and we all do at times), see if you can figure out where you went wrong.

I hope you've enjoyed this book. If you have any questions or comments, you can reach me at rainorshine@swiftpassage.com. Reviews are welcome.

About the Author

Ann Beardsley was a weather observer/forecaster for the Air National Guard in Texas, North Carolina, and Virginia for 12 years, and trained at the former Chanute Air Base in Rantoul, Illinois. Her husband, Elliott Walsh, is a private pilot with a lifelong interest in weather. (This makes birthdays and Christmas very easy.) They live in coastal Georgia where, like everywhere else, "if you don't like the weather, wait five minutes."

Printed in Great Britain
by Amazon

33695634R00052